S0-EQR-470

THE LETTER OF COLUMBUS

THE LETTER OF COLUMBUS
ON HIS DISCOVERY OF THE NEW WORLD

USC FINE ARTS PRESS

LOS ANGELES 1989

Copyright © 1989 by the University of Southern California

ISBN 0-944585-01-9

LC 88-21730

[EDITORIAL NOTE]

The facsimile of the Stephen Plannck 1493 Latin edition of *The Letter* was reproduced in actual size from photoengravings of the copy housed in Special Collections of Doheny Memorial Library.

Samuel Eliot Morison's English translation of *The Letter* is from his book *Christopher Columbus, Mariner,* and is used with the permission of Little, Brown and Company.

The woodcuts reproduced throughout this book are from the first illustrated edition of *The Letter,* the Basel Latin edition of 1493. These are thought to be based on designs sketched by Columbus himself. However, there is some evidence that several may have been taken from cuts printed in central Europe at an earlier date. The multiple and repeated use of illustrative matter by early printers was not uncommon.

The map reproduced on the title page is a sketch chart by Columbus of the northwest coast of Hispaniola, made during the first voyage. This sketch has also been attributed to Columbus' brother Bartholomew, although the original was among two folios of Christopher's manuscripts acquired by the Duchess of Alba about 1900. The cross-like symbol at the top of the map is thought to be unique to Columbus, possibly a secret sign to his son.

Oceanica Classis

[TABLE OF CONTENTS]

Introduction
by DOYCE B. NUNIS, JR., Professor of History,
University of Southern California,
and CHARLES R. RITCHESON, University Librarian
and Lovell University Professor, University of Southern California
[*page* 9]

Facsimile of the Stephen Plannck 1493 Latin Edition of The Letter
[*page* 21]

English Translation of The Letter
by Columbus historian SAMUEL ELIOT MORISON
[*page* 31]

Bibliographic Afterword
by LYNN F. SIPE, Assistant University Librarian,
University of Southern California
[*page* 41]

Introduction by DOYCE B. NUNIS, JR. and CHARLES R. RITCHESON

THE 1992 QUINCENTENNIAL of the European discovery of the New World celebrates one of the most significant anniversaries in the history of mankind. At two o'clock the morning of October 12, 1492, a lookout's cry signaled that a large section of the Earth's surface hitherto *incognita* had become *cognita*. The physical environment open to settlement and exploitation by the European peoples expanded immediately and enormously. In an astonishingly short time, unprecedented change began to penetrate to the profoundest levels of life, at first in the West, and ultimately, around the World.

The concern here is focused on the earliest stage of the change, its very inception, the discovery itself. The thrust into the unkown originated in Renaissance Italy, homeland of Christopher Columbus, the "Admiral of the Ocean Sea," and one of the greatest navigators in the annals of maritime enterprises, and in Spain, whose political and financial support undergirded the great voyage of 1492, and much of the age of discovery and settlement which followed. On the western side of the Atlantic was the *Novo Mundo,* the newly discovered lands comprising today the Caribbean nations, much of South and Central America, Mexico, and the United States whose national legend begins with Christopher Columbus, not Eric the Red, or Leif, his son.

Columbus' letter to the Court of Spain reporting what he had found on the first of his four voyages thus stands as one of the most portentous documents in human history. The University of Southern California Library proudly makes generally available the reprint of its own rare copy to mark the 500th anniversary of the great discovery.

In an extremely important sense, the discovery of America was one of history's greatest accidents. When he left European waters in

1492 sailing westward, Columbus was making for Cipangu (Japan) which, he was firmly convinced, lay some 3,000 miles across the Atlantic. The reason for this curious misapprehension lay in a letter written in 1474, by Pozzo Toscanelli, a respected Florentine physician and scholar of mathematics and astronomy. Responding to questions posed by a Lisbon geographer about the possibility of reaching the East by sailing west, Toscanelli accepted the assertion made in the 14th century by the Venetian Marco Polo that Europeans could reach Cathay (China) and Cipangu (Japan), the island lying off its shores, by sailing west through the Atlantic. A chart was enclosed indicating that Cipangu lay a mere 3,000 miles from the Canary Islands, and that the voyage thither would be free and clear from any serious physical hindrance. The theory was duly incorporated into the work of Martin Behaim of Nuremburg, the most celebrated globe maker of his day, who computed that the westward passage to Cipangu was precisely 3,080 miles from Europe.

Behaim made his great globe in 1492, too late for Columbus to see; but a copy of Toscanelli's earlier letter he obtained during a Lisbon visit in 1474 provided the fundamental basis for his geographical assumptions. When Columbus sailed out of the Spanish harbor of Palos on August 4, 1492, with a small flotilla of three caravels, the *Niña,* the *Pinta,* and *Santa María,* manned by ninety seamen, Columbus firmly believed they would arrive in the Far East after a comparatively short voyage.

The voyage was carefully chronicled by Columbus himself who kept a daily journal which he presented to Ferdinand and Isabella after the voyage. Those monarchs in turn placed the manuscript in the custody of the historian Bartolome de las Casas for safekeeping. The original has unfortunately disappeared, but Las Casas made an abridgment which survives to this day in the Archives of Simancas. It was finally published early in the 19th century by Martin Fernandez de Navarrete in his *Coleccion de los viages* (5 vols., Madrid, 1825-1837). John B. Thatcher translated the abridgment into English and published it again in his *Christopher Columbus* (2 vols., New York, 1903-1904). A more recent and critical edition, prepared by Oliver Dunn and James E. Kelley, Jr., has recently issued from the University of Oklahoma Press (Norman, 1988). With its sound scholarship and superb new translation of Las Casas, it clearly supercedes the versions by Fernandez de Navarrete and Thatcher. Since the original source is missing, Las Casas provides the sole source for tracing the course of Columbus' great voyage.

Having left the European continent behind, the flotilla made for Gomera in the Canary Islands on August 12th. Water and provisions were taken aboard. Becalmed for a time, the ships left harbor on September 9th. Due to a false landfall reported by Captain Martín Alonso Pinzón of the *Pinta*, course was altered to the southwest. (Had Columbus held to his original plan, he would probably have landed on the east coast of Florida.) At two a.m. on October 12th, the white cliffs of an island straight ahead were sighted glimmering in the moonlight; thus western history was divided into two parts. (Footnote: The date, October 12, 1492, is based on the Julian Calendar in use at the time. Under today's Gregorian Calendar, however, the anniversary becomes October 21st. Even so, the accepted date to commemorate Columbus Day remains October 12th, the only use made in the United States today of the Julian Calendar.)

Samuel Eliot Morison categorically asserted in his Pulitzer Prize winning biography of Columbus, *Admiral of the Ocean Sea* (Boston, 1944), that the initial New World landfall was Watlings Island in the

Bahamas, a conclusion he substantiated in the most direct way by himself sailing the course Las Casas described and making that landfall. Columbus called the place San Salvador in honor of Jesus Christ the Saviour, and the name was officially affixed in 1926. In November, 1986, the *National Geographic,* basing its case on computer evidence, declared in favor of Samana Cay, southeast of San Salvador a full degree of latitude in the Bahama Islands, as Columbus' true landfall. The discovery date itself remains unchallenged.

After a passage of thirty-three days and 3,409 miles (which squared with Toscanelli's chart), and believing himself in the vicinity of Cipangu, Columbus landed on the newly named San Salvador, took possession in the name of the Spanish monarchs—the first act of sovereignty in the New World—and explored the small island. Three days later, he visited a second island, which he named Santa María de la Concepción. On October 16th, a third island, "Fernandina," was discovered; and three days after that, a fourth was named Isabella. Columbus then sailed south, fully expecting to encounter the larger islands of Cipangu.

Cuba was reached on October 28th and named in honor of Juana, only child of Ferdinand and Isabella. Exploration of the northeast coast occupied the next nine days. On December 5th, Haiti was sighted and christened Española. Cruising along the northern coast, the small fleet suffered a major catastrophe: *Santa María,* the flagship, ran aground on a coral reef in Caracol Bay shortly after midnight on Christmas Day. Efforts to save the vessel, caught fast on the reef and battered mercilessly by rolling swells, proved unavailing; and, at daybreak, Columbus ordered the ship abandoned and transferred his flag to the *Niña,* which was standing by. A frantic salvage effort, assisted by friendly natives, retrieved much timber and fastenings from the hulk, which were put to use in building the first European structure in the New World, a fort appropriately called "Navidad."

On January 16, 1493, the *Niña* and *Pinta* headed eastward for the return voyage to Spain. Sometime in February, Columbus began to write an account of his discoveries for his patrons, Ferdinand and Isabella. A second "letter" was made as well to serve as a public announcement. On the 12th of the month, foul weather was encountered off the Azores; and in the midst of a dreadful storm which raged for the next two days, Columbus, fearing his ship would founder and all information about his discoveries would be lost, wrote a second account of the voyage. This account he wrapped first in parchment and then in waxed cloth. He had it sealed in a large wooden barrel and cast into the sea. This desperate measure proved fruitless; the barrel was never seen again.

At the height of the storm, on February 13th, the *Niña* and *Pinta* separated. On the 17th, the *Niña* reached safety at Santa Maria in the Azores. On the 24th, after a much-needed respite, the small vessel departed for the Iberian peninsula. Again, the weather proved savage. A cyclone on the 26th was succeeded by a bone-chilling cold front. Finally, at sunset on March 3rd, calm weather returned, and the *Niña* was saved by good luck and Columbus' superb seamanship. At daybreak, Cape Roco was sighted by the Roca da Sintra, a mountainous landmark of the Portuguese coast; and Columbus realized he was near the mouth of the channel which led to Lisbon. On Monday, March 4th, at 9 a.m., the *Niña* came to anchor at Restello (today's Belem), a short distance southwest of the Portuguese capital.

To King João II, Columbus made formal application for permission to come ashore; and to Ferdinand and Isabella, he dispatched both the letter written in February and the discovery announcement.

These he covered with another communication addressed to the Spanish sovereigns then with the royal court at Barcelona. He had made Lisbon, he wrote, "today, which was the greatest wonder in the world." The weather was understandably much on his mind. In "all the Indies" it had been May-like. (He) had gone there in thirty-three days, "and had returned in twenty-eight, but for those tempests which detained me fourteen days running through this sea. All mariners here say that never has there been so bad a winter or so many losses of ships."

In Lisbon, King João received Columbus a number of times, much interested in the account of the voyage and clearly determined to protect Portugal's pre-eminence in the exploration of the west coast of Africa. Old friends in navigational and geographical circles, beginning to understand that an extraordinary event of the first magnitude had occurred, lionized the discoverer.

On the morning of March 13th, the *Niña* weighed anchor and made for Spain. Shortly after midday two days later, she came to rest in the Rio Tinto, the harbor at Palos, the original port of embarkation. A few hours later, by the sheerest of coincidences, Columbus' second vessel, the *Pinta,* under Captain Martín Alonso, anchored near the *Niña*. Separated from the flagship at the height of the storm on February 13th, the *Pinta* had made Bayona, near Vigo, just north of the Spanish-Portuguese border about 450 miles from Palos, and had actually reached land before Columbus in the *Niña*. Whatever plans Alonso may have entertained for contesting with Columbus for the honor of first reporting the great discovery soon became moot. He died almost immediately, it is said, of chagrin.

Safely on Spanish soil, Columbus dispatched a courier via Seville with yet another copy of his letter and enclosure sent earlier to Ferdinand and Isabella from Lisbon. By March 19th, news of the discovery was in Madrid. Shortly thereafter, the sovereigns were fully informed in Barcelona, and immediately ordered copies of Columbus' announcement circulated among the court. (A number of manuscript copies still exist.) On March 30th, the royal couple acknowledged Columbus' dispatches.

This, then, is the background of *The Columbus Letter,* the "press release" written at sea on the *Niña* in February, 1493, announcing the great discovery to the literate public. The original and its accompanying letter to Ferdinand and Isabella has been lost; but there are two contemporary 1493-printed texts—one in Spanish, the other in Latin—in addition to a few manuscript copies. The Spanish printed version was prepared by Luis de Santagel (Santix), a member of the royal household. Probably the copyists and translators made some editorial changes; and it was rather badly printed on April 1st by Pedro Posa of Barcelona. A folio edition, the only known copy, is in the New York Public Library. A quarto Spanish edition appeared later in 1493, assumed to have been printed in Spain. A single surviving copy is to be found in the Biblioteca Ambrosiana in Milan.

News spread quickly. A Barcelona merchant, Hannibal Zenaro (Januarius), came into possession of either a manuscript copy or, perhaps, the Posa printing. On April 9th, he passed the information to his brother in Milan. It was from this brother that Jacome Trotti, the envoy to Milan from Ercole d'Este, the Duke of Ferrara, learned of the great development; and soon the Duke was demanding more details. On May 10th, Trotti complied, enclosing in his response a copy of what could only have been *The Columbus Letter* that had arrived from Spain.

About the middle of April, a copy of *The Columbus Letter* reached Rome; and on the 27th, word passed from the Eternal City to the Milanese envoy at Venice who immediately forwarded a copy to his lord, Ludovico il Moro. Thence, the news spread like wildfire throughout northern Italy, accelerated by a Latin translation from the Spanish made by a Catalan, Leandro de Cosco, and printed in Rome by Stephanus Plannck in May. Entitled *De Insulis inuentis. Epistola Cristoferi Colom,* it appeared as a newsletter or pamphlet of eight pages. This Latin version was widely republished and appeared under a variety of titles often chosen at the mere whim of the printer.

The Letter went through seventeen printings by 1497. In 1493, there were the first two editions in Spanish and the first edition and reprint by Plannck, and a third edition by Eucharius Argenteus (Silber). Bergmann de Olpe printed the Latin text in Basel, and a year later, in 1494, brought out a second printing. His 1493 edition contained pictorial illustrations, woodblock prints that served as decoration and actually had little if any relation to the contents of *The Letter.* In Paris, Guyot Marchand published three editions, although there is some doubt that all were printed in 1493, some scholars holding that two of the editions belong more properly to 1494. Thierry Martens published an Antwerp edition in 1494. The Florentine poet Giuliano Dati translated *The Letter* into Tuscan verse under the title *Questa e la hystoria della inuentioe delle diese Isole di Cannaria Indiane,* and had it printed in Florence on October 25, 1493. Under varying titles, the Italian version was reprinted that same year in Rome and twice again in Florence. Two additional Florentine editions followed in 1495. Finally, a German translation was printed in Strasburg by Bartholomew Küstler in 1497; and a second Spanish language printing by Giraldi was brought out at Valladolid, probably in 1497 as well.

Approximately seventy of these pre-1500 editions still exist, most of them being the Latin translation. The rarest copies are in Spanish:

Barcelona (1493) and Valladolid (c.1497). There is a single Antwerp-printed copy, and seven of Marchand's three Latin versions published in Paris in 1493 or 1494. The two Latin texts printed in Rome by Plannck have survived in larger number, some thirty-five copies in all. Also common is a Basel printing of 1494, curiously appended to a drama of the day written by Carlos Verardus about King Ferdinand's capture of Grenada in 1492 that is actually a reprint of Olpe's 1493 edition, including illustrations.

The very small number of the early printed editions of *The Columbus Letter* has meant a steadily rising market value for copies which rarely find their way to the world's auction houses or dealers. It means, too, that forgers have been active over the years. On the eve of the Quartracentennial, for example, a London publisher announced to startled scholars around the world that a Permbokeshire sailor had been given a trunk by a shipmate that contained a curious box holding "a ole buk." It was, the sailor told the Londoner, "rote on in ole leters an got pitchers in it thez ar ships as wud nevr sale an we carnt tel wat it is." He had consulted an expert mariner, the sailor continued, who said that the book was very old, belonged to Columbus, and was in fact, the very log recounting the voyage which discovered "mmerika." The "buk" was produced in stylish facsimile, lithographed on vellum, bound in sheepskin, covered with glued-on barnacles, sea shells, and seaweed—all materials imitation, to be sure—and offered to credulous souvenir hunters under the alluring title "My secrete Log Boke." Much of the oeuvre was based on *The Letter,* but recast in a Victorian rendering of Elizabethan prose. It is not known how many of these products found buyers at the Columbia Exposition in Chicago in 1893; but caveat emptor is a maxim as useful in the period of the Quincentennial as it would have been a century ago.

The facsimile edition here offered is from the first Latin text printed in Rome by Plannck in 1493. It belongs to the Special Collections of the University of Southern California Library, and is accompanied by a modern English translation and by the illustrations from the 1493 Basel Latin edition printed by Olpe. As a felicitous coincidence, it is also the first imprint of the University of Southern California's Fine Arts Press which finds its home in the University Library.

Fernanda
hyspania
ysabella
Saluatorie
Conceptoris
marie

Facsimile of the Stephen Plannck 1493 Latin edition of The Letter

¶ Epistola Christofori Colom: cui etas nostra multũ debet: de Insulis Indię supra Gangem nuper inuentis. Ad quas perquirendas octauo antea mense auspicijs τ ęre inuictissimi Fernandi Hispaniarum Regis missus fuerat: ad Magnificum dnm Raphaelem Sanxis: eiusdem serenissimi Regis Tesaurariũ missa: quam nobilis ac litteratus vir Aliander de Cosco ab Hispano Ideomate in latinum conuertit : tertio kal's Maij. M.cccc.xciij. Pontificatus Alexandri Sexti Anno primo.

Quoniam susceptę prouintię rem perfectam me cõsecutum fuisse gratum tibi fore scio: has constitui exarare: quę te vniuscuiusque rei in hoc nostro itinere gestę inuentęque ad moneant: Tricesimotertio die postq̃ Gadibus discessi in mare Indicũ perueni: vbi plurimas insulas innumeris habitatas hominibus repperi: quarum omnium pro foelicissimo Rege nostro pręconio celebrato τ vexillis extensis contradicente nemine possessionem accepi: primęque earum diui Saluatoris nomen imposui: cuius fretus auxilio tam ad hanc: q̃ ad cęteras alias peruenimus. Eam v̄o Indi Guanahanin vocant. Aliarum etiã vnam quanq̃ nouo nomine nuncupaui. Quippe aliã insulam Sanctę Marię Conceptionis. aliam Fernandinam . aliam Hysabellam. aliam Johanam. τ sic de reliquis appellari iussi. Quamprimum in eam insulam quã dudum Johanã vocari dixi appulimus: iuxta eius littus occidentem versus aliquantulum processi: tamq̃ eam magnã nullo reperto fine inueni: vt non insulam: sed continentem Chatai prouinciam esse crediderim: nulla tñ videns oppida municipiaue in maritimis sita confinibꝰ pręter aliquos vicos τ prędia rustica: cum quoꝝ incolis loqui nequibam quare simul ac nos videbant surripiebant fugam. Progrediebar vltra: existimans aliquã me vrbem villasue inuenturum. Deniq̃ videns q̃ longe admodum progressis nihil noui emergebat: τ hmõi via nos ad Septentrionem deferebat: q̃ ipse fugere exoptabã: terris etenim regnabat bruma: ad Austrumq̃ erat in voto cõtendere

nec minus venti flagitantibꝰ succedebāt · cōstitui alios nō ope
riri successus: ꞇ sic retrocedens ad portū quendā quem signaue/
ram sum reuersus: vnde duos boies ex nostris in terrā misi. qui
inuestigarēt esset ne Rex in ea prouincia vrbesue aliquę. Di per
tres dies ambularunt inuenerūtꝙ inumeros populos ꞇ habita
tiones paruas tn̄ ꞇ absꝙ vllo regimine: quapropter redierunt·
Interea ego iam intellexerā a qbusdam Indis quos ibidem su/
sceperā quō hm̄oi prouincia insula quidem erat: ꞇ sic perrexi ori
entem versus eius semp stringēs littora vsꝙ ad miliaria·cccxxij
vbi ipsius insulę sunt extrema: hinc aliā insulam ad orientē pro
spexi distantē ab hac Johana miliaribꝰ·liiij·quā protinus Dispa
nam dixi: in eamꝗ concessi ꞇ direxi iter quasi per Septentrionē
quemadmodū in Johana ad orientem miliaria·dlxiiij·que dicta
Johana ꞇ alię ibidem insulę ꝙfertilissime existunt· Hęc multis
atꝙ tutissimis ꞇ latis nec aliis quos vnꝙ viderim comparādis
portibus est circundata multi maximi ꞇ salubres banc interflu
unt fluuij multi quoꝗ ꞇ eminentissimi in ea sunt montes Ōes
hę insulę sunt pulcherrimę ꞇ varijs distinctę figuris: puię: ꞇ ma
xima arboꝝ varietate sidera lambentiū plenę: quas nunꝙ folijs
priuari credo Quippe vidi eas ita virentes atꝙ decoras: ccu mē
se Maio in Dispaniā solent esse: quaꝝ alię florētes: alię fructuo/
sę: alię in alio statu bm̄ vniuscuiusꝙ qualitatē vigebant · garrie/
bat philomena ꞇ alij passeres vari ac inumeri mēse Nouembris
quo ipse per eas deambulabā · Sunt pterea in dicta insula Joha
na septē vel octo palmaꝝ genera ꝗ proceritate ꞇ pulchritudine
queadmodū cętere ōes arbores: herbe: fructusꝙ nr̄as facile exu/
perāt · Sūt ꞇ mirabiles pinꝰ agri ꞇ prata vastissima·varię aues·
varia mella·variaꝙ metalla ferro excepto· In ea aūt quā Dispa
nam supra diximꝰ nuncupari maximi sunt mōtes ac pulchri. va
sta rura nemora·campi feracissimi seri pascuꝗ ꞇ xdendis ędifici
is aptissimi·Portuū in hac insula cōmoditas ꞇ prestantia·flumi
nū copia salubritate admixta boim̄: q̄ nisi quis viderit: credulita
tez superat·Huiꝰ arbores pascua ꞇ fructꝰ multū ab illis Johanę
differunt Hęc preterea Dispana diuerso aromatis genere·auro.

metalliscp abundat·cuius quidem τ oium alia㈜ quas ego vidi τ
qua㈜ cognitionẽ habeo incolę vtriuscp sexus nudi semper ince/
dunt queadmodũ edunt in lucem:pręter aliquas feminas:q̃ fo/
lio frondeue aliqua aut bombicino velo pudenda operiunt· qd̃
ipsę sibi ad id negocii parant· Carent ij oẽs(vt supra dixi)quo/
cũcp genere ferri·carẽt τ armis vpote sibi ignotis nec ad ea sunt
apti·nõ ppter corporis deformitatem cũ sint bene formati: sed
qa sunt timidi ac pleni formidine·gestant tñ pro armis arundi/
nes sole pustas:in qua㈜ radicib? hastile quoddã ligneũ siccũ et
in mucronem attenuatũ figunt:neq̃ ijs audent iugiter vti:nã sę
pe euenit cũ miserim duos vel tris hoies ex meis ad aliquas vil/
las vt cũ ea㈜ loquerent incolis:exi·sse agmen glomeratũ ex Jn
dis:τ vbi nr̃os appropinquare videbant fugã celeriter arripuis/
se despretis a patre liberis τ econtra·τ hoc nõ cp cuipiam eo㈜ dã
num aliqd̃ vel iniuria illata fuerit: imo ad quoscũcp appuli τ q/
bus cũ verbum facere potui:quicqd habebã sum elargitus: pan
num alia㈜ pmultu nulla mihi facta versura:sed sunt natura pa
uidi ac timidi·Cete㈜ vbi se cernunt tutos oĩ metu repulso: sunt
admodũ simplices ac bonę fidei τ in oib? que habent liberalissi/
mi:roganti qd̃ possidet inficiat̃ nemo:quin ipsi nos ad id poscẽ
dũ inuitãt·Maximũ erga oẽs amorẽ preseferunt:dant qcq̃ ma
gna pro paruis:minima lz re nihiloue 𝔢tenti·ego attñ phibui ne
tã minia τ nulli? precii hisce darent̃: vt sunt lancis paraphsidũ·
vitriq̃ fragmẽta·ite claui ligulę:quanq̃ si hoc poterãt adipisci
videbat̃ eis pulcherrima mũdi possidere iocalia·Accidit·n·quẽ
dã nauitam tantũ auri pondus habuisse pro vna ligula quati
sunt tres aurei solidi·τ sic alios pro alijs minoris precii·psertim
pro blanquis nouis:quibusdã nũmis aureis: p qb? habẽdis da
bant quicquid petebat vẽditor:puta vnciam cũ dimidia τ duas
auri:vel triginta τ quadraginta bombicis pondo: quam ipsi ia㈜
nouerant·item arcuum·amphorę·hydre·doliiq̃ fragmenta bom
bice τ auro tanq̃ bestię comparabant· quod quia iniquum sane
erat vetui:dediq̃ eis multa pulchra τ grata q̃ mecũ tulerã nullo
ĩteruenĩete pr̃mio:vt eos mihi facili? 𝔠iliarem fierentq̃ xr̃picolę

τ vt sint proni in amorem erga Regem Reginã principemqz noſtros τ vniuerſas gentes Hiſpaniẹ ac ſtudeãt perquirere:coaceruare eaqz nobis tradere quibᵘ ipſi affluunt τ nos magnopere indigemus. Nullam ij norunt hydolatriam:imo firmiſſime credũt omnem vim:oẽm potentiam:oĩa deniqz bona eſſe in coelo:meqz inde cum his nauibus τ nautis deſcẽdiſſe: atqz hoc animo vbiqz fui ſuſceptus poſtqz metum repulerant. Hec ſunt ſegnes aut rudes:quin ſummi ac perſpicacis ingenij:τ homines qui transfretant mare illud nõ ſine admiratõe vniuſcuiuſqz rei ratioñe reddunt:ſed nunqz viderũt gentes veſtitas neqz naues hmõi. Ego ſtatim atqz ad mare illud perueni e prima inſula quoſdã Indos violenter arripui:qui ediſcerent a nobis τ nos pariter docerent ea:quoꝝ ipſi in iſce partibus cognitionem habebant . τ ex voto ſucceſſit:nam breui nos ipſos τ ij nos tum geſtu ac ſignis tũ verbis intellexerunt:magnoqz nobis fuere emolumento: veniũt modo mecum:qui ſemper putant me deſiluiſſe e coelo: Quis diu nobiſcum verſati fuerint hodieqz verſentur . τ ij erant primi qui id quocunqz appellabamus nuntiabant: alij deinceps alijs elata voce diẽtẽes: Venite venite τ videbitis gentes ethereas. Quã ob rem tam feminẹ qz viri: tam impuberes qz adulti: tã iuuenes qz ſenes depoſita formidine paulo ante ꝯcepta nos certatim viſebant magna iter ſtipante caterua alijs cibum alijs potum afferentibus maximo cum amore ac beniuolentia incredibili. Habet vnaquẹqz inſula multas ſcaphas ſolidi ligni:τ ſi anguſtas longitudine tñ ac forma noſtris biremibus ſimiles: curſu aũt velociores. Reguntur remis tantumodo. Haꝝ quedaz ſunt magnẹ: quedam paruẹ:quedã in medio conſiſtũt. Plures tñ biremi quẹ remiget duodeuiginti tranſtris maiores: cũ quibus in oẽs illas inſulas:quẹ innumerẹ ſunt:traijcitur.cumqz his ſuam mercaturam exercent τ inter eos comertia fiunt. Aliquas ego harum biremiũ ſeu ſcapharꝝ vidi q̃ vehebant ſeptuaginta τ octuaginta remiges. In omnibus ijs inſulis nulla eſt diuerſitas inter gentis effigies :nulla in moribus atqz loquela: quin oẽs ſe intelligunt adinuicem:quę res perutilis est ad id qd serenissimũ Regẽ nr̃m exoptare pręcipue reor:scz̃ eoȝ ad sctãm xp̃i fidem ꝯuersionem. cui qdem quantũ intelligere potui facillimi sunt ꞇ proni. Dixi quemadmodũ sum progressus antea insulam Johanã per rectũ tramitem occasus in orientem miliaria·ccccxxij· ĩ quã viam ꞇ interuallum itineris possum dicere hanc Johanã esse maiorem Anglia ꞇ Scotia simul:nãcɜ vltra dicta·ccccxxij· passuũ milia in ea parte quę ad occidentem prospectat duę: quas nõ petij:super sunt prouincię:quaȝ alterã Indi Anan vocant cuius accolę cau dati nascuntur·Tendunt̃ in longitudineȝ ad miliaria·clxxx·vt ab his quos veho mecũ Indis percepi:qui ois has callent insu las·Dispanę ꝟo ambitꝯ maior est tota Dispania a Colonia vscɜ ad fontem rabidum hinccɜ facile arguit̃ q̃ quartum eius latus quod ipse per rectã lineã occidentis in orientem traieci miliaria continet·dxl·Dęc insula est affectanda ꞇ affectata nõ spernenda In qua ꞇ si aliaȝ oim vt dixi pro inuictissimo Rege nostro solen niter possessionem accepi:earũcɜ imperium dicto Regi penitus cõmittitur:in oportuniori tñ loco atcɜ omni lucro et cõmercio condecenti cuiusdã magnę villę:cui Natiuitatis dñi nomen de dimus·possessionem peculiariter accepi:ibicɜ arcem quandam erigere extemplo iussi:quę modo iam debet esse pacta:in qua ho mines qui necessarij sunt visi cũ omni armoȝ genere ꞇ vltra an num victu oportuno reliqui·Ite quandã carauellam ꞇ pro alijs construendis tam in hac arte q̃ in ceteris peritos:ac eiusdẽ in sulę Regis erga eos beniuolentiam ꞇ familiaritatem incredibilẽ Sunt enim gentes illę amabiles admodum ꞇ benignę:eo q̃ Rex predictus me fratrẽ suũ dici gloriabat̃·Et si aĩum reuocarent et ijs qui in arce manserunt nocere velint:nequeunt:qa armis ca rent:nudi incedũt ꞇ nimiũ timidi:ideo dictã arcem tenẽtes dun taxat p̃nt totã eam insulam nullo sibi iminẽte discrimine popu lari·dummõ leges quas dedimꝯ ac regimen nõ excedãt·In oibꝯ ijs insulis vt intellexi quiscɜ vni tm̃ ꝯiugi aqcuiescit preter prin cipes aut reges:qbus viginti bie licet·Feminę magis q̃ viri la

borare videntur·nec bene poſſui intelligere an habeant bona pro
pria:vidi enim qɔ vnus habebat alijs impartiri:pręſertim dapes
obſonia τ bmōi·Nullum apud eos monſtrũ reperi vt pleriqɔ exi
ſtimabant:ſed hoīes magnę reuerentię atqɔ benignos·Hec ſunt
nigri velut ethiopes·habent crines planos ac demiſſos· non de
gũt vbi radioɤ ſolaris emicat calor·pmagna nãcɔ hic eſt ſolis
vehementia:propterea qɔ ab ęquinoctiali linea diſtat.Ubi vide
tur gradus ſex τ viginti ex montiũ cacuminibɔ·Maximũ quocɔ
viget frigus:ſed id qdem moderant Jndi tum loci conſuetudi
ne·tum rerũ calidiſſimaɤ quibɔ frequenter τ luxurioſe veſcunt
pręſidio·Jtacɔ mõſtra aliqua nõ vidi:necɔ eoɤ alicubi habui co
gnitionem:excepta quadã inſula Charis nuncupata:quę ſecun
da ex Hiſpania in Jndiam tranſfretantibɔ exiſtit·quã gens quę
dam a finitimis habita ferociox incolit·Hi carne humana veſcũ
tur·Habent predicti biremiũ genera plurima qbus in ois Jndi
cas inſulas traijciunt·deprędant.ſurripiunt quęcũcɔ pñt·Nihil
ab alijs differunt niſi qɔ gerũt more femineo longos crines vtũ
tur arcubɔ τ ſpiculis arundineis fixis vt dirimɔ in groſſiori par
te attenuatis haſtilibɔ·ideocɔ habent feroces:quare ceteri Jndi
inexhauſto metu plectunt:ſed hos ego nihilifacio plus cɔ̃ alios
Hi ſunt q coheunt cũ quibuſdã feminis:quę ſolę inſulã Mateu
nin primã ex Hiſpania in Jndiã traijcientibɔ habitant· Hę aũt
feminę nullũ ſui ſexus opus exercent: vtuntur enim arcubus et
ſpiculis ſicuti de eaɤ ɔiugibus dixi muniunt ſeſe laminis ęneis
quaɤ maxima apud eas copia exiſtit.Aliã mihi inſulã affirmãt
ſupradicta Hiſpana maiore:eius incolę carẽt pilis·aurocɔ inter
alias potiſſimũ exuberat·Huius inſulę τ aliaɤ quas vidi hoīes
mecũ porto qui hoɤ quę dixi teſtimoniũ perhibẽt·Denicɔ vt no
ſtri diſceſſus τ celeris reuerſionis compendiũ ac emolumentũ
breuibus aſtringã hoc polliceor:me noſtris Regibus inuictiſſi
mis parno eoɤ fultũ auxilio:tantũ auri daturũ quantũ eis fue
rit opus·tm̃ vero aromatum·bombicis·maſticis:cɔ̃ apud Chium
duntaxat inuenitur:tantũcɔ lignũ aloes· tantum ſeruoɤ hydro

latroꝛum: quantum eoꝛum maleſtas voluerit exigere. Item reu
barbarum ⁊ alia aromatū genera quę iȷ quos in dicta arce reli
qui iam inueniſſe atqȝ inuenturos exiſtimo. qñquidem ego nul
libi magis ſum moꝛatus niſi quantum me coegerunt venti: pꝛe
terqȝ in villa Natiuitatis dum arcem condere ⁊ tuta oīa eſſe pꝛo
uidi · Quae ⁊ ſi maxima ⁊ inaudita ſunt: multo tñ maioꝛa foꝛẽt
ſi naues mihi vt ratio exigit ſubueniſſent. Vep multum ac mira
bile hoc: nec noſtris meritis coꝛreſpondens: ſed ſanctę Chꝛiſtia
nę fidei noſtrorumqȝ Regum pietati ac religioni: quia quod hu
manus conſequi nō poterat intellectus: id humanis cōceſſit dí
uinus. Solet enim deus ſeruos ſuos quiqȝ ſua pꝛęcepta diligunt
⁊ in impoſſibilibus exaudire: vt nobis in pꝛęſentia contigit: qui
ea conſecuti ſumus quę hactęnus mortalium vires minime atti
gerant: nam ſi harū inſulaꝝ quipiam aliquid ſcripſerunt aut lo
cuti ſunt: omnes per ambages ⁊ cōiecturas: nemo ſe eas vidiſſe
aſſerit vnde pꝛoꝑe videbatur fabula. Igitur Rex ⁊ Regina pꝛin
cepſqȝ ac eoꝛ regna feliciſſima cunctęqȝ aliȷ Chꝛiſtianoꝝ pꝛcuin
cię Saluatoꝛi dño noſtro Jeſu Chꝛiſto agamᵒ gratias: qui tan
ta nos victoꝛia munereqȝ donauit: celebꝛentur proceſſiones: per
agantur ſolennia ſacra. feſtaqȝ fronde velentur delubꝛa. exultet
Chꝛiſtus in terris quemadmodum in cœlis exultat: quom tot
populoꝛ perditas ante hac animas ſaluatum iri pꝛęuidet. Lęte
mur ⁊ nos: cum pꝛopter exaltationem noſtrę fidei. tum pꝛopter
rerum temporalium incrementa: quoꝝ non ſolum Hiſpania ſed
vniuerſa Chꝛiſtianitas eſt futura particeps.   Hęc vt geſta ſunt
ſic bꝛeuiter enarrata. Uale. Uliſbonę pꝛidie idus Martij.

    Chꝛiſtofoꝛus Colom Oceanę claſſis Pꝛęfectus.

¶ Epigramma · R · L · de Corbaria Episcopi Montispalusij·
Ad Inuictissimum Regem Hispaniarum.

Jam nulla Hispanis tellus addenda triumphis
  Atq̃ parum tantis viribus orbis erat·
Nunc longe eois regio deprensa sub vndis
  Auctura est titulos Betice magne tuos
Vnde repertori merito referenda Columbo
  Gratia:sed summo est maior habenda deo·
Qui vincenda parat noua regna tibiq̃ sibiq̃
  Teq̃ simul fortem prestat et esse pium.

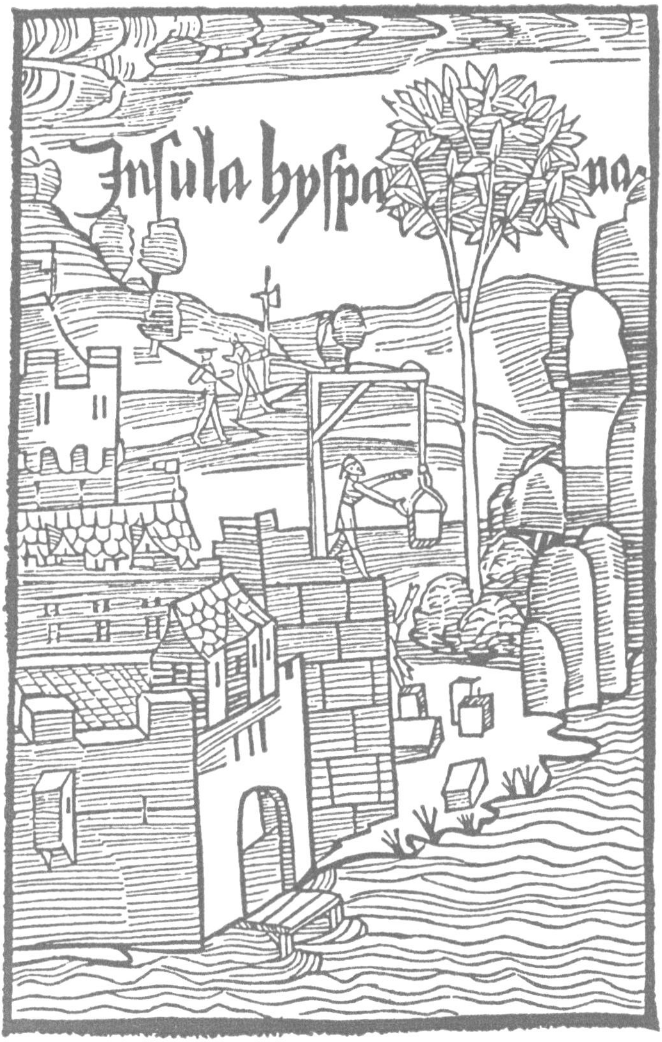
Insula hyspana

English Translation of The Letter by SAMUEL ELIOT MORISON

Sir, since I know that you will take pleasure at the great victory with which Our Lord has crowned my voyage, I write this to you, from which you will learn how in twenty days I reached the Indies with the fleet which the most illustrious King and Queen, our lords, gave to me. And there I found very many islands filled with people without number, and of them all I have taken possession for their Highnesses, by proclamation and with the royal standard displayed, and nobody objected. To the first island which I found I gave the name *Sant Salvador,* in remembrance of His Heavenly Majesty, who marvelously hath given all this; the Indians call it *Guanahani.* To the second I gave the name *Isla de Santa Maria de Concepción;* to the third, *Ferrandina;* to the fourth, *La Isla Bella;* to the fifth, *La Isla Juana;* and so to each one I gave a new name.

When I reached Juana, I followed its coast to the westward, and I found it to be so long that I thought it must be the mainland, the province of Catayo. And since there were neither towns nor cities on the coast, but only small villages, with the people of which I could not have speech because they all fled forthwith, I went forward on the same course, thinking that I should not fail to find great cities and towns. And, at the end of many leagues, seeing that there was no change and that the coast was bearing me to the north, which was contrary to my desire since winter was already beginning and I proposed to go thence to the south, and as moreover the wind was favorable, I determined not to wait for a change of weather and backtracked to a notable harbor; and thence I sent two men upcountry to learn if there were a king or great cities. They traveled for three days and found an infinite number of small villages and people without number, but nothing of importance; hence they returned.

I understood sufficiently from other Indians, whom I had already taken, that continually this land was an island, and so I followed its coast eastwards 107 leagues up to where it ended. And from that cape I saw toward the east another island, distant eighteen leagues from the former, to which I at once gave the name *La Spañola*. And I went there and followed its northern part, as I had in the case of Juana, to the eastward for 178 great leagues in a straight line. As Juana, so all the others are very fertile to an excessive degree, and this one especially. In it there are many harbors on the coast of the sea, incomparable to others which I know in Christendom, and numerous rivers, good and large, which is marvelous. Its lands are lofty and in it there are very many sierras and very high mountains, to which the island *Centrefrei* is not comparable. All are most beautiful, of a thousand shapes, and all accessible and filled with trees of a thousand kinds and tall, and they seem to touch the sky; and I am told that they never lose their foliage, which I can believe, for I saw them as green and beautiful as they are in Spain in May, and some of them were flowering, some with fruit, and some in another condition according to their quality. And there were singing the nightingale and other little birds of a thousand kinds in the month of November, there where I went. There are palm trees of six or eight kinds, which are a wonder to behold on account of their beautiful variety, and so are the other trees and fruits and herbs; therein are marvelous pine groves, and extensive champaign country; and there is honey, and there are many kinds of birds and a great variety of fruits. Upcountry there are many mines of metals, and the population is innumerable. *La Spañola* is marvelous, the sierras and the mountains and the plains and the champaigns and the lands are so beautiful and fat for planting and sowing, and for livestock of every sort, and for building towns and cities. The harbors of the sea here are such as you could not believe in without seeing them, and so the rivers, many and great, and good streams, the most of which bear gold. And the trees and fruits and plants have great differences from those of La Juana; in this there are many spices and great mines of gold and of other metals.

The people of this island and of all the other islands which I have found and seen, or have not seen, all go naked, men and women, as their mothers bore them, except that some women cover one place only with the leaf of a plant or with a net of cotton which they make for that. They have no iron or steel or weapons, nor are they capable of using them, although they are well-built people of handsome stature, because they are wonderfully timorous. They have no other arms than arms of canes, [cut] when they are in seed time, to the ends of which they fix a sharp little stick; and they dare not make use of these, for oftentimes it has happened that I have sent ashore two or three men to some town to have speech and, people without number have come out to them, and as soon as they saw them coming, they fled; even a father would not stay for his son; and this not because wrong has been done to anyone; on the contrary, at every point where I have been and have been able to have speech, I have given them of all that I had, such as cloth and many other things, without receiving anything for it; but they are like that, timid beyond cure. It is true that after they have been reassured and have lost this fear, they are so artless and so free with all they possess, that no one would believe it without having seen it. Of anything they have, if you ask them for it, they never say no; rather they invite the person to share it, and show as much love as if they were giving their hearts; and whether the thing be of value or of small price, at once they are content with whatever little thing of whatever kind may be given to them. I forbade that they should be given things so worthless as pieces of broken crockery and broken glass, and ends of straps, although when they were able to get them, they thought they had the best jewel in the world; thus it was ascertained that a sailor for a strap received gold to the weight of two and a half *castellanos,* and others much more for other things which were worth much less; yea, for new *blancas,* for them they would give all that they had, although it might be two or three castellanos' weight of gold or an *arrova* or two of spun cotton; they even took pieces of the broken hoops of the wine casks and, like animals, gave what they had, so that it seemed to me to be wrong and I forbade it, and I gave them a thousand good, pleasing things which I had brought, in order that they might be fond of us, and furthermore might be made Christians and be inclined to the love and service of their Highnesses and of the whole Castilian nation, and try to help us and to give us of the things which they have in abundance and which are necessary to us. And they know neither sect nor idolatry, with the exception that all believe that the source of all power and goodness is in the sky, and they believe very firmly that I, with these ships and people, came from the sky, and in this belief they everywhere received me, after they had overcome their fear. And this does not result from their being ignorant, for they are of a very keen intelligence and men who navigate all those seas, so that it is marvelous the good account they give of everything, but because they have never seen people clothed or ships like ours.

And as soon as I arrived in the Indies, in the first island which I found, I took by force some of them in order that they might learn [Castilian] and give me information of what they had in those parts; it so worked out that they soon understood us, and we them, either by speech or signs, and they have been very serviceable. I still have them with me, and they are still of the opinion that I come from the sky, in spite of all the intercourse which they had with me, and they were the first to announce this wherever I went, and the others went running from house to house and to the neighboring towns with loud cries of, "Come! Come! See the people from the sky!" Then all came, men and women, as soon as they had confidence in us, so that not one, big or little, remained behind, and all brought something to eat and drink, which they gave with marvelous love. In all the islands they have very many *canoas* like rowing *fustes,* some bigger and some smaller, and some are bigger than a *fusta* of eighteen benches. They are not so broad, because they are made of a single log, but a *fusta* could not keep up with them by rowing, since they make incredible speed, and in these [canoes] they navigate all those islands, which are innumerable, and carry their merchandise. Some of these canoes I have seen with seventy and eighty men in them, each one with his oar.

In all these islands, I saw no great diversity in the appearance of the people or in their manners and language, but they all understand one another, which is a very singular thing, on account of which I hope that their Highnessess will determine upon their conversion our holy faith, towards which they are much inclined.

I have already said how I went 107 leagues in a straight line from west to east along the coast of the island Juana, and as a result of that voyage I can say that this island is larger than England and Scotland together; for, beyond these 107 leagues, there remain to the westward two provinces where I have not been, one of which they call *Auau,* and there the people are born with tails. Those provinces cannot have a length of less than fifty or sixty leagues, as I could understand from those Indians whom I retain and who know all the islands. The other, *Española,* in circuit is greater than all Spain, from *Colunya* by the coast to *Fuenterauia* in Vizcaya, since I went along one side 188 great leagues in a straight line from west to east. It is a desirable land and, once seen, is never to be relinquished; and in it, although of all I have taken possession for their Highnesses and all are more richly supplied than I know or could tell, I hold them all for their Highnesses, which they may dispose of as absolutely as of the realms of Castile. In this *Española,* in the most convenient place and in the best district for the gold mines and for every trade both with this continent and with that over there belonging to the *Gran Can* [Grand Khan], where there will be great trade and profit, I have taken possession of a large town to which I gave the name *La Villa de Navidad,* and in it I have built a fort and defenses, which ready, at this moment, will be all complete, and I have left in it enough people for such a purpose, with arms and artillery and provisions for more than a year, and a *fusta,* and a master of the sea in all arts to build others; and great friendship with the king of that land, to such an extent that he took pride in calling me and treating me as brother; and even if he were to change his mind and offer insult to these people, neither he nor his know the use of arms and they go naked, as I have already said, and are the most timid people in the world, so that merely the people whom I have left there could destroy all that land; and the island is without danger for their persons, if they know how to behave themselves.

In all these islands, it appears, all the men are content with one woman, but to their *Maioral,* or king, they give up to twenty. It appears to me that the women work more than the men. I have been unable to learn whether they hold private property, but it appeared true to me that all took a share in anything that one had, especially in victuals.

In these islands I have so far found no human monstrosities, as many expected; on the contrary, among all these people good looks are esteemed; nor are they Negroes, as in Guinea, but with flowing hair, and they are not born where there is excessive force in the solar rays; it is true that the sun there has great strength, although it is distant from the Equator twenty-six degrees. In these islands, where there are high mountains, the cold this winter was strong, but they endure it through habit and with the help of food which they eat with many and excessively hot spices. Thus I have neither found monsters nor had report of any, except in an island which is the second at the entrance to the Indies, which is inhabited by a people who are regarded in all the islands as very ferocious and who eat human flesh; they have many canoes with which they range all the islands of India and pillage and take as much as they can; they are no more malformed than the others, except that they have the custom of wearing their hair long like women, and they use bows and arrows of the same stems of a cane with a little piece of wood at the tip for want of iron, which they have not. They are ferocious toward these other people, who are exceeding great cowards, but I make no more account of them than of the rest. These are those who have intercourse with the women of *Matremonio,* which is the first island met on the way from Spain to the Indies, in which there is not one man. These women use no feminine exercises, but bows and arrows of cane, like the abovesaid; and they arm and cover themselves with plates of copper, of which they have plenty. In another island, which they assure me is larger than *Española,* the people have no hair. In this there is countless gold, and from it and from the other islands I bring with me Indios as evidence.

In conclusion, to speak only of that which has been accomplished on this voyage, which was so hurried, their Highnesses can see that I shall give them as much gold as they want if their Highnesses will render me a little help; besides spice and cotton, as much as their Highnesses shall command; and gum mastic, as much as they shall order shipped, and which up to now, has been found only in Greece, in the island of Chios, and the Seignory sells it for what it pleases;

and aloe wood, as much as they shall order shipped, and slaves, as many as they shall order, who will be idolaters. And I believe that I have found rhubarb and cinnamon, and I shall find a thousand other things of value, which the people whom I have left there will have discovered, for I have not delayed anywhere, provided the wind allowed me to sail, except in the town of Navidad, where I stayed [to have it] secured and well seated. And the truth is I should have done much more if the ships had served me as the occasion required.

This is sufficient. And the eternal God, Our Lord, Who gives to all those who walk in His way victory over things which appear impossible, and this was notably one. For although men have talked or have written of these lands, all was conjecture, without getting a look at it, but amounted only to this, that those who heard for the most part listened and judged it more a fable than that there was anything in it, however small.

So, since our Redeemer has given this victory to our most illustrious King and Queen, and to their famous realms, in so great a matter, for this all Christendom ought to feel joyful and make great celebrations and give solemn thanks to the Holy Trinity with many solemn prayers for the great exaltation which it will have, in the turning of so many peoples to our holy faith, and afterwards for material benefits, since not only Spain but all Christians will hence have refreshment and profit. This is exactly what has been done, though in brief.

Done in the caravel, off the Canary Islands, on the fifteenth of February, year 1493. At your service.

THE ADMIRAL

Insula hyspana

Bibliographic Afterword by LYNN F. SIPE

THE COPY of *The Columbus Letter* reprinted in this volume came to the University of Southern California Library in early 1926 through the generosity of the University's President at the time, Rufus B. von KleinSmid. He purchased it from the London firm of Maggs Bros. who offered it in their Catalog #470 (Christmas 1925) at a price of £1,250.

Little more is known about the provenance of this particular copy of *The Letter* but a great deal is known about the edition itself and its relationship to the other early editions of *The Letter*. The history of the bibliographic knowledge and study of *The Columbus Letter* provides an interesting sub-text to the much larger discussion of *The Letter's* content and significance. This Bibliographic Afterword acknowledges, in summary fashion, the most significant contributions in the evolution of *Columbus Letter* scholarship.

"The original of the letter of Columbus, describing the general results of his first voyage, is not known to be now in existence. Several versions of it, however, have been preserved, and while no one of them can be regarded as an exact copy of the original, it is possible from them to reconstruct with some approach to certainty the text of the lost document."[1] These words by Cecil Jane suggest the direction most scholarly bibliographic attention to *The Columbus Letter* has taken: identification, textual comparison and analysis of the early (15th century) editions of *The Letter,* and related consideration of the historical context and other collateral factors which might shed light on this fascinating and complex subject.

Most of the historical, analytical, and descriptive bibliography of *The Letter* is to be found in certain of the scholarly historical studies of Columbus, in bibliographic essays on *The Letter,* and in annotated bibliographic listings, including published library catalogs. In addition, much valuable analytical and descriptive comment is included with many of the facsimile printings of one or more of the early editions, or with reprintings and translations of *The Letter's* text.

The existence of all nineteen editions of *The Letter* was, of course, not always a matter of common scholarly knowledge. The first extended enumeration of editions of *The Letter* listed but ten of the early printings.[2] The first edition of Graesse listed only seven of the incunabula editions.[3] Henry Harrisse provided the first extended scholarly attention to the bibliography of *The Columbus Letter* as part of his meticulous and monumental *Bibliotheca Americana Ventustissima* (1866, with a separate set of additions issued in 1872).[4] This work is not only of fundamental importance in providing the first detailed descriptions of *The Letter* and its context, as known at that time, but is also a monument in the field of early American bibliography as well. Harrisse provides extremely complete physical descriptions of all editions known to him, including their provenance.

Between the first appearance of Harrisse's work and his subsequent additions a significant contribution to the bibliography of *The Letter* was published by the Hakluyt Society under the editorship of R. H. Major.[5] Major's collection includes seven letters of Columbus, including one version of *The Columbus Letter*. More important is its inclusion of an extended bibliographic essay focusing on the six editions of *The Letter* known to Major. This work makes no reference to Harrisse's pioneering effort. Major's bibliography was subsequently published separately in a very limited edition of seventy-five copies.[6]

Much useful information regarding *The Letter* is found in the "Note on Columbus" in the Ticknor Catalog of 1879.[7] A chapter in Harrisse's major documentary study of Columbus, published in

1884/1885, is devoted to propagation of the news of the discovery, achieved initially through printing of *The Letter*.[8] A lengthy set of Notes to Justin Winsor's *Narrative and Critical History of America* provides valuable comparisons of the differing attributions of priority of seven editions of *The Letter* along with a wealth of scholarly detail.[9] The 400th anniversary of Columbus' voyage brought forth, as might be expected, a flurry of publications dealing with the discovery and *The Letter*. Most significant among these is de Lollis's *Scritti di Cristoforo Colombo*.[10] The author's meticulous attention to detail and exhaustive scholarly documentation certainly make this one of the landmarks in *Columbus Letter* scholarship.

Another essay from Harrisse appeared in 1892, published under the pseudonymn of B.A.V.[11] This interesting article focuses, in much detail, on rebutting a theory advanced previously that the Ambrosian Library copy of *The Letter* was published in Seville instead of the now accepted location of Valladolid.

The most complete nineteenth century bibliography in the Spanish language was compiled by the pre-eminent Latin American bibliographer Jose Toribio Medina in 1898 as part of his larger compilation on Hispanic-American history.[12] It enumerates sixteen of the nineteen 15th century editions of *The Letter* with detailed analysis, description, and comparison. Extensive secondary references are also provided.

John Boyd Thacher provided the first major twentieth century contribution in *Columbus Letter* scholarship in his monumental three volume study of Columbus.[13] In his chapters on *The Letter* Thacher pays particular attention to textual and typographic variations between the major editions. He appends a very useful table identifying and distinguishing between each of the 15th century editions.

Another good summary of basic bibliographic details for editions of *The Letter* is given in the famous catalog of the Church Library.[14] This collection was purchased for the Huntington Library in 1911. Another useful listing of a library's holdings is that issued by the John Carter Brown Library at Brown University.[15] The Brown catalog provides excellent descriptions of their six incunabula editions of *The Letter* as well as of a number of facsimiles in their possession. Finally, an important article by the famous American bibliographer, Wilberforce Eames, published in 1924, includes a frequently cited listing of the various editions of *The Letter*.[16]

Cecil Jane's edition of documents on the voyages of Columbus for the Haklyut Society includes extensive notes on *The Letter*.[17] Jane relies very heavily on the work of de Lollis in his analysis. Two articles, also by Jane, appeared the same year as his Haklyut publication (1930), one focusing on *The Letter* and the other on the question of Columbus's literacy.[18] Both provide interesting and useful commentary.

A useful compilation for locating copies of *The Letter* in the holding libraries is the standard catalog for incunabula, the *Gesamtkatalog der Wiegendrucke*.[19]

The Spanish historian Carlos Sanz is to recent years what Henry Harrisse was to his time, the day's pre-eminent bibliographic scholar of *The Columbus Letter*. Sanz published a major historical essay on *The Letter* in 1957, resulting from a conference at the Spanish National Library on the occasion of the 465th anniversary of Columbus's first voyage.[20] He is also responsible for the invaluable, definitive bibliography on *The Columbus Letter*[21] issued in 1958. The bibliography, containing 420 citations, is extensively and critically annotated. The author has listed all relevant books, articles, essays, and significant early historical references to *The Columbus Letter* from 1493 through mid-1958. Sanz also published a major bibliographical study of the Latin translation of *The Letter* in 1959.[22]

A very useful popular account of *The Letter* was published in 1971 by Maurice Lubin in *Americas,* the magazine of the Organization of American States.[23] S.R.Wilson provides a unique perspective on *The Columbus Letter* in his 1978 article on its literary qualities.[24]

The first facsimile printing of an edition of *The Columbus Letter* was done by Adam Pilinski in Paris in 1858.[25] His choice was an incomplete copy of the Latin version issued in Basel in 1493. Since Pilinski's initial effort there have been several facsimile editions of various editions of *The Letter.* With but three significant exceptions noted below, our purpose here is not to enumerate these but to acknowledge that the current edition merely continues in this tradition.[26]

The first exception is the only other fine press (letterpress-printed) facsimile edition of which we are aware, which was published by Edwin Grabhorn in 1924.[27] There is also another important contribution, again from Carlos Sanz, in editing a portfolio of facsimiles of the first seventeen printed editions.[28] Finally, we must note and admire the recently published facsimile of the only known copy of the Barcelona edition of the Spanish version of *The Letter,* from the holdings of the New York Public Library.[29] This elaborate and expensive edition is not only exceptionally attractive, particularly in its use of color reproduction, but also provides valuable collateral materials, including three cartographic facsimiles.

[NOTES]

Listed in brackets after each citation is the item number from the definitive Sanz bibliography on *The Columbus Letter* (see Note 21), for works published before 1958. Sanz provides detailed bibliographic citations and often lengthy, valuable annotations.

1. Lionel Cecil Jane. *Select documents illustrating the four voyages of Columbus, including those contained in R. H. Major's Select letters of Christopher Columbus.* Translated and edited with additional material, an introduction and notes by Cecil Jane. London: Printed for the Hakluyt Society, 1930, p. cxxiii. [Sanz #355]

2. James Lenox. "Bibliographical notice of the early accounts of Columbus voyages" in his printing of *Nicolaus Syllacius de Insulis meridiani atque Indice maris nuper inventis.* With a translation into English by the Rev. John Mulligan, A. M. New York, 1859, pp. xxxv-lxii. [Sanz #150]

3. Jean George Theodore Grasse. *Tresor de livres rares et precieux, ou nouveau Dictionnaire Bibliographique....* Dresden: Rudolf Kuntze, Libraire-Editeur, 1859-1869. [Sanz #149]

4. Henry Harrisse. *Bibliotheca Americana Ventustissima; a description of works relating to America, published between 1492 and 1551.* New York: George P. Philes, Publisher, 1866, pp. 1-34. [Sanz #171] The additions were published as *A description of works relating to America, published between the years, 1492-1551. Additions.* Paris: Tross, 1872. [Sanz #183] The original edition, with the Additions, has had three reprintings—Madrid: Libreria General V. Suarez, 1958, edited by Carlos Sanz; Chicago: Argonaut, 1967; and Amsterdam: P. Schippers, 1967.

5. Christopher Columbus. *Select letters of Christopher Columbus, with other original documents, relating to his four voyages to the New World.* Translated and edited by R.H. Major. 2nd edition. London: Printed for the Hakluyt Society, 1870, pp. v-lxxxix. [Sanz #179] The first edition was printed for the Society in 1847 [Sanz #130] with considerably less bibliographical detail included.

6. Richard Henry Major. *The Bibliography of the first letter of Columbus, describing his discovery of the New World.* London: llis & White, 1872. [Sanz #184] Printed in a limited edition of 75 copies. Reprinted Amsterdam: Meridian Publishing Co., 1971.

7. James Lyman Whitney. "Notes on Columbus" in *Catalogue of the Spanish library and of the Portuguese books bequeathed by George Ticknor to the Boston Public Library....* Boston: Printed by Order of the Trustees, 1879, pp. 92-95. [Not in Sanz]

8. Henry Harrisse. *Christophe Colomb; son origine, sa vie, ses voyages, sa famille et ses descendents....* Paris: E. Leroux, 1884, pp. 10-42. [Sanz #210]

9. Justin Winsor, ed. *Narrative and critical history of America*. Boston: Houghton, Mifflin and Company, 1866, vol. 2, pp. 46-51. [Sanz #174] A second edition was published in 1886. [Sanz #220]

10. Cesare de Lollis. *Scritti di Cristoforo Colombo*. Rome: R. Commissione Colombiana, 1892, parte I, volumnen I, pp. xxv-lxxiv; 120-135. [Sanz #262]

11. B. A. V. (Henry Harrisse). "Qui a imprime la premiere lettre de Colomb?" *Centralblatt fur Bibliothekswesen* (Leipzig) 9, no. 3 (March 1892): 105-122. [Sanz #257] Also published as a separate pamphlet, Leipzig: Harrassowitz, 1892.

12. Jose Toribio Medina. *Biblioteca Hispano-Americana, 1493-1810*. Santiago de Chile: Printed by the author, 1898. [Sanz #295]

13. John Boyd Thacher. *Christopher Columbus: his life, his work, his remains, as revealed by original printed and manuscript records....* New York: G. P. Putnam's sons, 1903, vol. 2, pp. 3-72. [Sanz #309]

14. Elihu Dwight Church. *A Catalogue of books relating to the discovery and early history of North and South America, forming a part of the library of E. D. Church*. Compiled and annotated by George Watson Cole. New York: Dodd, Mead and company, 1907, vol. 1, pp. 6-11. [Sanz #316] Reprinted New York: Peter Smith, 1951.

15. Brown University. John Carter Brown Library. *Bibliotheca Americana; catalogue of the John Carter Brown Library in Brown University, Providence, Rhode Island*. Providence: Published by the Library, 1919, vol. 1, pp. 16-21. [Sanz #324] Reprinted Nendeln, Liechtenstein: Kraus Reprint Corporation, 1961.

16. Wilberforce Eames. "Two important gifts by Mr. George F. Baker, Jr. I. Columbus' Letter on the Discovery of America (1493-1497)," *Bulletin of the New York Public Library*, 28, no. 8 (August 1924): 595-599. [Sanz #339]

17. Lionel Cecil Jane. *Select documents...., op. cit.*, vol. 1, pp. cxxiii-cxliii. Reprinted Nendeln, Liechtenstein: Kraus Reprint Limited, 1967.

18. *Ibid*. "The letter of Columbus announcing the success of his first voyage." *The Hispanic American Historical Review*, 10, no. 1 (1930): 33-50. [Sanz #356] and "Notes and Comment: the question of the literacy of Columbus in 1492," *The Hispanic American Historical Review*, 10, no. 4 (1930): pp. 500-516. [Not in Sanz]

19. *Gesamtkatalog der Wiegendrucke*. Leipzig: Verlag von Karl W. Hiersemann, 1934, band VI, pp. 758-762. [Not in Sanz]

20. Carlos Sanz. *'La Carta de Colon' anunciando la llegada a las Indias (descubrimineto de America); critica historica*. Madrid: 1957. [Sanz #415]

21. Ibid. *Bibliografia general de la carta de Colon*. Madrid: Libreria General Victoriano Suarez, 1958.

22. Ibid. *El gran secreto de la Carta de Colon (critica historica) y otras adiciones a la Bibliotheca Americana Vetustissima*. Madrid: Libreria General Victoriano Suarez. 1959.

23. Maurice A. Lubin. "First letter from the New World." *Americas*, 23, no. 4 (April 1971): 2-12. Reprinted in *Manuscripts*, 23, no. 4 (Fall 1971): 244-256.

24. S. R. Wilson. "The Form of discovery: the Columbus letter announcing the finding of America." *Revista Canadiense de Estudios Hispanicos* (Toronto) 2, no. 2 (Winter 1978): 154-168.

25. Paris: A. Pilinski. 1858. Facsimile of *De Insulis inuentis Epistola Cristoferi Colom*.... Basle: Bergman de Olpe o Jacof Wolff de Prorzheim, 1493. [Sanz #146]

26. The most complete listings of facsimiles can be found in the John Carter Brown Catalog (no. 15) and the Sanz bibliography (no. 21).

27. Christopher Columbus. *The Letter of Christopher Columbus concerning his first voyage to the New World*. Done into English & provided with a foreword by Donald B. Clarck. San Francisco: Printed by E. Grabhorn, 1924. [Sanz #337]

28. Cristoforo Colombo. *La carta de Colon anunciando la llegada a las Indias y a la Provincia de Catayo, China; descubrimiento de America. Reproduccion facsimilar de las 17 ediciones conocidas*. Madrid: 1958.

29. *The Columbus Papers; the Barcelona Letter of 1493*.... New York/Barcelona: The New York Public Library and NYW Publishing, Ltd., 1987.

Fernādº rex hyspania

[PRODUCTION NOTES]

The Letter of Columbus was designed and produced under the direction of Gerald Lange at the USC Fine Arts Press. The book was hand printed at the Press by Robin Price. The typefaces we have used are English Monotype Poliphilus and its companion italic Blado. The type was cast by Harold Berliner's Typefoundry in Nevada City, California, and reset by hand at the Press. The paper is Frankfurt Cream, a German mould-made from the Zerkall mill. The book was bound by Klaus-Ullrich S. Rötzscher in San Francisco. The photoengravings were produced at Peterson Engraving Company of Los Angeles. The edition consists of twenty-six copies lettered from A through Z and 300 numbered copies.

This is copy

127